Houghton
Mifflin
Harcourt

Made in the United States
Text printed on 100%
recycled paper

Houghton Mifflin Harcourt

Printed in the U.S.A.

ISBN 978-0-544-34202-6

6 7 8 9 10 0928 22 21 20 19 18 17 16 15

4500540873   B C D E F G

Dear Students and Families,

Welcome to **Go Math!**, Grade 2! In this exciting mathematics program, there are hands-on activities to do and real-world problems to solve. Best of all, you will write your ideas and answers right in your book. In **Go Math!**, writing and drawing on the pages helps you think deeply about what you are learning, and you will really understand math!

By the way, all of the pages in your **Go Math!** book are made using recycled paper. We wanted you to know that you can Go Green with **Go Math!**

Sincerely,

The Authors

Made in the United States
Text printed on 100% recycled paper

GO MATH!

Authors

Juli K. Dixon, Ph.D.
Professor, Mathematics Education
University of Central Florida
Orlando, Florida

Edward B. Burger, Ph.D.
President, Southwestern University
Georgetown, Texas

Steven J. Leinwand
Principal Research Analyst
American Institutes for
    Research (AIR)
Washington, D.C.

Contributor

Rena Petrello
Professor, Mathematics
Moorpark College
Moorpark, California

Matthew R. Larson, Ph.D.
K-12 Curriculum Specialist for
    Mathematics
Lincoln Public Schools
Lincoln, Nebraska

Martha E. Sandoval-Martinez
Math Instructor
El Camino College
Torrance, California

English Language Learners Consultant

Elizabeth Jiménez
CEO, GEMAS Consulting
Professional Expert on English
    Learner Education
Bilingual Education and
    Dual Language
Pomona, California

Addition and Subtraction

Critical Area Building fluency with addition and subtraction

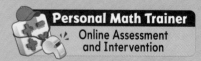
Critical Area

 6 **3-Digit Addition and Subtraction**   **387**

COMMON CORE STATE STANDARDS

2.NBT Number and Operations in Base Ten
Cluster B Use place value understanding and properties of operations to add and subtract.
2.NBT.B.7, 2.NBT.B.9

GO DIGITAL

Go online! Your math lessons are interactive. Use *iTools*, Animated Math Models, the Multimedia *eGlossary*, and more.

Chapter 6 Overview

In this chapter, you will explore and discover answers to the following **Essential Questions**:

- What are some strategies for adding and subtracting 3-digit numbers?
- What are the steps when finding the sum in a 3-digit addition problem?
- What are the steps when finding the difference in a 3-digit subtraction problem?
- When do you need to regroup?

Personal Math Trainer
Online Assessment and Intervention

CRITICAL AREA REVIEW PROJECT PLAN A TRIP TO THE ZOO: *www.thinkcentral.com*

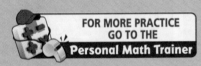

FOR MORE PRACTICE
GO TO THE
Personal Math Trainer

Practice and Homework

Lesson Check and
Spiral Review in
every lesson

Chapter 6

3-Digit Addition and Subtraction

Curious about Math

Monarch butterflies roost together during migration.

If you count 83 butterflies on one tree and 72 on another, how many butterflies have you counted altogether?

Name _____

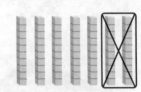
Model Subtracting Tens

Write the difference. (1.NBT.C.6)

1.

5 tens — 3 tens = _____ tens

50 — 30 = _____

2.

7 tens — 2 tens = _____ tens

70 — 20 = _____

2-Digit Addition

Write the sum. (2.NBT.B.6)

3. $\begin{array}{r} 54 \\ + 25 \\ \hline \end{array}$ 4. $\begin{array}{r} 35 \\ + 18 \\ \hline \end{array}$ 5. $\begin{array}{r} 82 \\ + 67 \\ \hline \end{array}$ 6. $\begin{array}{r} 29 \\ + 81 \\ \hline \end{array}$

Hundreds, Tens, and Ones

Write the hundreds, tens, and ones shown. Write the number. (2.NBT.A.1)

7.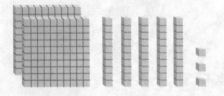

Hundreds	Tens	Ones

_____

8.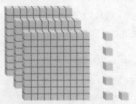

Hundreds	Tens	Ones

_____

This page checks understanding of important skills needed
for success in Chapter 6.

Vocabulary Builder

Review Words

regroup
sum
difference
hundreds

Visualize It

Fill in the graphic organizer by writing examples of ways to **regroup**.

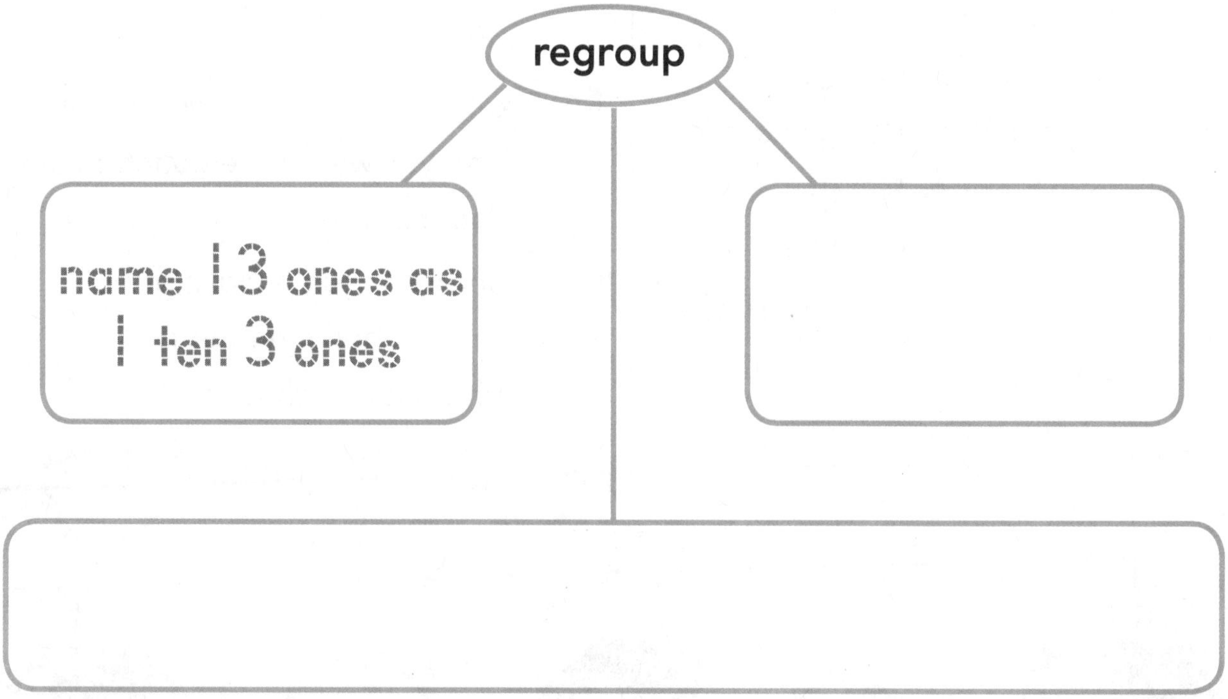

regroup

name 13 ones as
1 ten 3 ones

Understand Vocabulary

1. Write a number that has a **hundreds** digit that is greater than its tens digit. _____

2. Write an addition sentence that has a **sum** of 20. _____

3. Write a subtraction sentence that has a **difference** of 10. _____

GO DIGITAL
• Interactive Student Edition
• Multimedia eGlossary

Game 2-Digit Shuffle

Materials
- number cards 10–50
- 15 • 15 ⬤

Play with a partner.

1. Shuffle the number cards. Place them face down in a pile.
2. Take two cards. Say the sum of the two numbers.
3. Your partner checks your sum.
4. If your sum is correct, place a counter on a button. If you regrouped to solve, place a counter on another button.
5. Take turns. Cover all the buttons. The player with more counters on the board wins.
6. Repeat the game, saying the difference between the two numbers for each turn.

Chapter 6 Vocabulary

addend

sumando

1

column

columna

7

difference

diferencia

14

digit

dígito

15

hundred

centena

31

is equal to (=)

es igual a

33

regroup

reagrupar

56

sum

suma o total

59

column

$$
\begin{array}{r}
3\,\boxed{3} \\
3\,\boxed{4} \\
+3\,\boxed{2}
\end{array}
$$

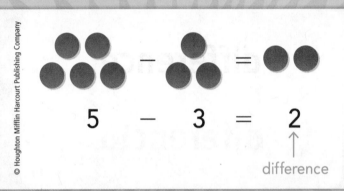

5 + 3 = 8

addends

0, 1, 2, 3, 4, 5, 6, 7, 8, and 9 are digits.

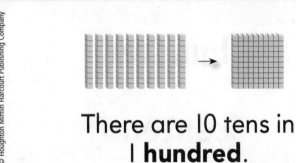

5 − 3 = 2

difference

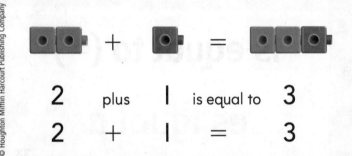

2   plus   1   is equal to   3

2 + 1 = 3

There are 10 tens in 1 **hundred**.

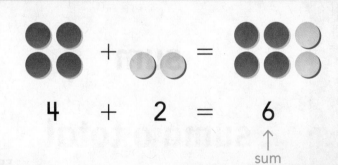

4 + 2 = 6

sum

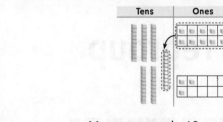

Tens	Ones

You can trade 10 ones for 1 ten to **regroup**.

Picture It

For 3 to 4 players

Word Box
addends
column
difference
digit
hundred
is equal to
regroup
sum

Materials
- timer
- sketch pad

How to Play

1. Take turns to play.

2. Choose a math word from the Word Box. Do not say the word.

3. Set the timer for 1 minute.

4. Draw pictures and numbers to give clues about the math word.

5. The first player to guess the word gets 1 point. If he or she use the word in a sentence, they get 1 more point. That player gets the next turn.

6. The first player to score 5 points wins.

The Write Way

Reflect

Choose one idea. Write about it on the lines below.

- Tell how to solve this problem.

$$42 - 25 = \underline{\hspace{2cm}}$$

- Write a paragraph that uses at least **three** of these words.

    addends   digit   sum   hundred   regroup

- Explain something you know about regrouping.

Name _____

Draw to Represent 3-Digit Addition

Essential Question How do you draw quick pictures to show adding 3-digit numbers?

Common Core — **Number and Operations in Base Ten—2.NBT.B.7**

MATHEMATICAL PRACTICES
MP2, MP5, MP6

Listen and Draw (Real World)

Draw quick pictures to model the problem.
Then solve.

Tens	Ones
	_____ pages

Math Talk

MATHEMATICAL PRACTICES 5

Use Tools Explain how your quick pictures show the problem.

FOR THE TEACHER • Read this problem to children. Manuel read 45 pages in a book. Then he read 31 more pages. How many pages did Manuel read? Have children draw quick pictures to solve the problem.

three hundred ninety-one **391**

Model and Draw

Add 234 and 141.

Hundreds	Tens	Ones

___3___ hundreds ___7___ tens ___5___ ones

___375___

Share and Show MATH BOARD

Draw quick pictures. Write how many hundreds, tens, and ones in all. Write the number.

✓ 1. Add 125 and 344.

Hundreds	Tens	Ones

_____ hundreds _____ tens _____ ones

_____

✓ 2. Add 307 and 251.

Hundreds	Tens	Ones

_____ hundreds _____ tens _____ ones

_____

Name _____

On Your Own

Draw quick pictures. Write how many hundreds, tens, and ones in all. Write the number.

3. Add 231 and 218.

Hundreds	Tens	Ones

_____ hundreds _____ tens _____ ones

_____

4. Add 232 and 150.

Hundreds	Tens	Ones

_____ hundreds _____ tens _____ ones

_____

5. **THINK SMARTER** Use the quick pictures to find the two numbers being added. Then write how many hundreds, tens, and ones in all. Write the number.

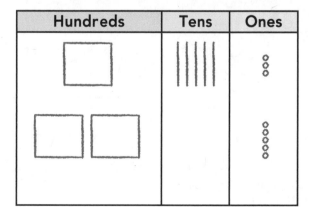

Add _____ and _____.

_____ hundreds _____ tens _____ ones

_____

Problem Solving • Applications

 **WRITE** Math

6. **MATHEMATICAL PRACTICE ②** **Represent a Problem**

There are 125 poems in Carrie's book and 143 poems in Angie's book. How many poems are in these two books?

Draw a quick picture to solve.

_____ poems

Personal Math Trainer

7. **THINK SMARTER +** Rhys wants to add 456 and 131.

Help Rhys solve this problem. Draw quick pictures. Write how many hundreds, tens, and ones in all. Write the number.

Hundreds	Tens	Ones

_____ hundreds _____ tens _____ ones

_____

 TAKE HOME ACTIVITY • Write 145 + 122. Have your child explain how he or she can draw quick pictures to find the sum.

Name _____

Draw to Represent 3-Digit Addition

Common Core **COMMON CORE STANDARD—2.NBT.B.7**
Use place value understanding and properties of operations to add and subtract.

Draw quick pictures. Write how many hundreds, tens, and ones in all. Write the number.

1. Add 142 and 215.

Hundreds	Tens	Ones

_____ hundreds _____ tens _____ ones

_____

Problem Solving

Solve. Write or draw to explain.

2. A farmer sold 324 lemons and 255 limes. How many pieces of fruit did the farmer sell altogether?

_____ pieces of fruit

3. WRITE Math Draw quick pictures and write to tell how you would add 342 and 416.

_____

_____

_____

Lesson Check (2.NBT.B.7)

1. Ms. Carol sold 346 child tickets and 253 adult tickets. How many tickets did Ms. Carol sell?

_____ tickets

2. Mr. Harris counted 227 gray pebbles and 341 white pebbles. How many pebbles did Mr. Harris count?

_____ pebbles

Spiral Review (2.OA.C.4, 2.NBT.B.5, 2.NBT.B.6)

3. Pat has 3 rows of shells. There are 4 shells in each row. How many shells does Pat have?

_____ shells

4. Kara counted 32 red pens, 25 blue pens, 7 black pens, and 24 green pens. How many pens did Kara count?

_____ pens

5. Kai had 46 blocks. He gave 39 blocks to his sister. How many blocks does Kai have left?

$46 - 39 =$ _____ blocks

6. A shop has 55 posters for sale. 34 posters show sports. The rest of the posters show animals. How many posters show animals?

_____ posters

FOR MORE PRACTICE
GO TO THE
Personal Math Trainer

Name _____

Break Apart 3-Digit Addends

Essential Question How do you break apart addends to add hundreds, tens, and then ones?

Common Core
Number and Operations in Base Ten—2.NBT.B.7
MATHEMATICAL PRACTICES
MP6, MP8

Listen and Draw

Write the number. Draw a quick picture for the number. Then write the number in different ways.

_____

_____ hundreds _____ tens _____ ones

_____ + _____ + _____

_____

_____ hundreds _____ tens _____ ones

_____ + _____ + _____

FOR THE TEACHER • Have children write 258 on the blank in the left corner of the first box. Have children draw a quick picture for this number and then complete the other two forms for the number. Repeat the activity for 325.

Math Talk

MATHEMATICAL PRACTICES 6

Make Connections
What number can be written as 400 + 20 + 9?

Chapter 6

Model and Draw

Break apart the addends into hundreds, tens, and ones. Add the hundreds, the tens, and the ones. Then find the total sum.

538 $\longrightarrow$ 500 + 30 + 8

+216 $\longrightarrow$ 200 + 10 + 6

700 + ___ + ___ = _____

Share and Show

Break apart the addends to find the sum.

1. 321 $\longrightarrow$ _____ + _____ + _____

    +457 $\longrightarrow$ _____ + _____ + _____

    _____ + _____ + _____ = _____

2. 744 $\longrightarrow$ _____ + _____ + _____

    +162 $\longrightarrow$ _____ + _____ + _____

    _____ + _____ + _____ = _____

3. 254 $\longrightarrow$ _____ + _____ + _____

    +536 $\longrightarrow$ _____ + _____ + _____

    _____ + _____ + _____ = _____

On Your Own

Break apart the addends to find the sum.

4.   374 ⟶ _____ + _____ + _____

    +518 ⟶ _____ + _____ + _____

         _____ + _____ + _____ = _____

5.   425 ⟶ _____ + _____ + _____

    +232 ⟶ _____ + _____ + _____

         _____ + _____ + _____ = _____

6.   849 ⟶ _____ + _____ + _____

    +123 ⟶ _____ + _____ + _____

         _____ + _____ + _____ = _____

7.   **THINK SMARTER** Mr. Jones has many sheets of paper. He has 158 sheets of blue paper, 100 sheets of red paper, and 231 sheets of green paper. How many sheets of paper does he have?

_____ sheets of paper

Problem Solving • Applications  Math

8. **GO DEEPER** Wesley added in a different way.

Use Wesley's way to find the sum.

```
  327
+ 468
```

```
  700      7 hundreds
   80      8 tens
+  15      15 ones
  795
```

```
  539
+ 247
```

9. **THINK SMARTER** There are 376 children at one school.
There are 316 children at another school.
How many children are at the two schools?

376 $\longrightarrow$ 300 + 70 + 6

+ 316 $\longrightarrow$ 300 + 10 + 6

Select one number from each column to solve the problem.

Hundreds	Tens	Ones
○ 2	○ 4	○ 2
○ 4	○ 8	○ 3
○ 6	○ 9	○ 6

 TAKE HOME ACTIVITY • Write 347 + 215. Have your child
break apart the numbers and then find the sum.

Break Apart 3-Digit Addends

Common Core COMMON CORE STANDARD—2.NBT.B.7
Use place value understanding and properties of operations to add and subtract.

Break apart the addends to find the sum.

1.   518 ⟶ _____ + _____ + _____

   + 221 ⟶ _____ + _____ + _____

   _____ + _____ + _____ = _____

2.   438 ⟶ _____ + _____ + _____

   + 142 ⟶ _____ + _____ + _____

   _____ + _____ + _____ = _____

Problem Solving

Solve. Write or draw to explain.

3. There are 126 crayons in a bucket.
   A teacher puts 144 more crayons
   in the bucket. How many crayons
   are in the bucket now?

   _____ crayons

4. **WRITE** Math Draw quick pictures
   and write to explain how to
   break apart addends to find
   the sum of 324 + 231.

_____

Lesson Check (2.NBT.B.7)

1. What is the sum?

$$\begin{array}{r} 218 \\ + 145 \\ \hline \end{array}$$

2. What is the sum?

$$\begin{array}{r} 664 \\ + 223 \\ \hline \end{array}$$

Spiral Review (2.OA.B.2, 2.NBT.B.5, 2.NBT.B.6, 2.NBT.B.9)

3. Ang found 19 berries and Barry found 21 berries. How many berries did they find altogether?

$19 + 21 = $ _____ berries

4. Write a subtraction fact related to $9 + 6 = $

_____

5. There are 25 goldfish and 33 betta fish. How many fish are there?

$25 + 33 = $ _____ fish

6. Subtract 16 from 41. Draw to show the regrouping. What is the difference?

Tens	Ones

_____

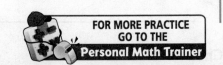

FOR MORE PRACTICE
GO TO THE
Personal Math Trainer

Name _____

3-Digit Addition: Regroup Ones

Essential Question When do you regroup ones in addition?

Common Core
Number and Operations in Base Ten—2.NBT.B.7
MATHEMATICAL PRACTICES
MP4, MP6, MP8

Listen and Draw Real World   Hands On

Use ▨ ▭ . to model the problem.
Draw quick pictures to show what you did.

Hundreds	Tens	Ones

FOR THE TEACHER • Read the following problem and have children model it with blocks. There were 213 people at the show on Friday and 156 people at the show on Saturday. How many people were at the show on the two nights? Have children draw quick pictures to show how they solved the problem.

Math Talk MATHEMATICAL PRACTICES 6

Describe how you modeled the problem.

Add the ones.

$6 + 7 = 13$

Regroup 13 ones as 1 ten 3 ones.

Hundreds	Tens	Ones
	1	
2	4	6
+ 1	1	7
		3

Hundreds	Tens	Ones

Add the tens.

$1 + 4 + 1 = 6$

Hundreds	Tens	Ones
	1	
2	4	6
+ 1	1	7
	6	3

Hundreds	Tens	Ones

Add the hundreds.

$2 + 1 = 3$

Hundreds	Tens	Ones
	1	
2	4	6
+ 1	1	7
3	6	3

Hundreds	Tens	Ones

Share and Show  MATH BOARD

Write the sum.

✓ 1.

Hundreds	Tens	Ones
3	2	8
+ 1	3	4

✓ 2.

Hundreds	Tens	Ones
4	4	5
+	2	3

On Your Own

Write the sum.

3.

Hundreds	Tens	Ones
	☐	
5	2	6
+ 1	0	3

4.

Hundreds	Tens	Ones
	☐	
3	4	8
+	1	9

5.

Hundreds	Tens	Ones
	☐	
6	2	8
+ 3	4	7

6.

Hundreds	Tens	Ones
	☐	
2	3	5
+ 2	5	7

7.

Hundreds	Tens	Ones
	☐	
5	6	2
+ 3	2	9

8.

Hundreds	Tens	Ones
	☐	
1	4	7
+ 1	2	5

9. **THINK SMARTER** On Thursday, there were 326 visitors at the zoo. There were 200 more visitors at the zoo on Friday than on Thursday. How many visitors were at the zoo on both days?

_____ visitors

Problem Solving • Applications  WRITE Math

Solve. Write or draw to explain.

10. **MATHEMATICAL PRACTICE ④ Model with Mathematics** The gift shop is 140 steps away from the zoo entrance. The train stop is 235 steps away from the gift shop. How many total steps is this?

_____ steps

11. **THINK SMARTER** Katina's class used 249 noodles to decorate their bulletin board. Gunter's class used 318 noodles. How many noodles did the two classes use?

_____ noodles

Did you have to regroup to solve? Explain.

 TAKE HOME ACTIVITY • Ask your child to explain why he or she regrouped in only some of the problems in this lesson.

3-Digit Addition: Regroup Ones

Common Core

COMMON CORE STANDARD—2.NBT.B.7
Use place value understanding and properties of operations to add and subtract.

Write the sum.

1.

Hundreds	Tens	Ones
	☐	
1	4	8
+ 2	3	4

2.

Hundreds	Tens	Ones
	☐	
3	2	1
+ 3	1	8

3.

Hundreds	Tens	Ones
	☐	
4	1	4
+ 1	7	9

4.

Hundreds	Tens	Ones
	☐	
6	0	2
+ 2	5	8

Problem Solving Real World

Solve. Write or draw to explain.

5. In the garden, there are 258 yellow daisies and 135 white daisies. How many daisies are in the garden altogether?

_____ daisies

6. **WRITE** Math Find the sum of 136 + 212. Explain why you did or did not regroup.

_____

Lesson Check (2.NBT.B.7)

I. What is the sum?

Hundreds	Tens	Ones
	□	
4	3	5
+ 1	4	6

2. What is the sum?

Hundreds	Tens	Ones
	□	
4	3	6
+ 3	0	6

Spiral Review (2.OA.B.2, 2.NBT.B.5, 2.NBT.B.6, 2.NBT.B.7)

3. What is the difference?

$$9 - 4 = \underline{\quad}$$

4. What is the sum?

$$\begin{array}{r} 82 \\ + 59 \\ \hline \end{array}$$

5. What is the sum?

$$26 + 7 = \underline{\quad}$$

6. Add 243 and 132. How many hundreds, tens, and ones are there in all?

____ hundreds ____ tens

____ ones

FOR MORE PRACTICE
GO TO THE
Personal Math Trainer

Name _____

3-Digit Addition: Regroup Tens

Essential Question When do you regroup tens in addition?

Common Core Number and Operations in Base Ten—2.NBT.B.7

MATHEMATICAL PRACTICES
MP6, MP8

Listen and Draw Real World   Hands On

Use  to model the problem.
Draw quick pictures to show what you did.

Hundreds	Tens	Ones

Math Talk

MATHEMATICAL PRACTICES 6

Explain how your quick pictures show what happened in the problem.

FOR THE TEACHER • Read the following problem and have children model it with blocks. On Monday, 253 children visited the zoo. On Tuesday, 324 children visited the zoo. How many children visited the zoo those two days? Have children draw quick pictures to show how they solved the problem.

Model and Draw

Add the ones.

$2 + 5 = 7$

Hundreds	Tens	Ones
☐	☐	
1	4	2
+ 2	8	5
		7

Hundreds	Tens	Ones

Add the tens.

$4 + 8 = 12$

Regroup 12 tens as
1 hundred 2 tens.

Hundreds	Tens	Ones
1	☐	
1	4	2
+ 2	8	5
	2	7

Hundreds	Tens	Ones

Add the hundreds.

$1 + 1 + 2 = 4$

Hundreds	Tens	Ones
1	☐	
1	4	2
+ 2	8	5
4	2	7

Hundreds	Tens	Ones

Share and Show

Write the sum.

1.

Hundreds	Tens	Ones
☐	☐	
3	4	7
+ 2	9	1

✓ 2.

Hundreds	Tens	Ones
☐	☐	
1	6	5
+ 3	5	4

✓ 3.

Hundreds	Tens	Ones
☐	☐	
5	3	8
+ 1	4	0

Name _____

Write the sum.

4.

Hundreds	Tens	Ones
☐	☐	
1	5	6
+	4	2

5.

Hundreds	Tens	Ones
☐	☐	
7	6	4
+ 1	5	3

6.

Hundreds	Tens	Ones
☐	☐	
3	7	2
+ 1	8	5

7.

2	2	4
+ 1	5	7

8.

3	1	4
+ 4	3	5

9.

7	5	3
+ 1	5	2

10. GO DEEPER In a bowling game, Jack scored 116 points and 124 points. Hal scored 128 points and 134 points. Who scored more points? How many more points were scored?

_____ _____ more points

MATHEMATICAL PRACTICE 6 Attend to Precision
Rewrite the numbers. Then add.

11. 760 + 178

+ _____

12. 216 + 346

+ _____

13. 423 + 285

+ _____

Problem Solving • Applications  WRITE Math

14. **THINK SMARTER** These lists show the pieces of fruit sold. How many pieces of fruit did Mr. Olson sell?

Mr. Olson	Mr. Lee
257 apples	314 pears
281 plums	229 peaches

_____ pieces of fruit

15. **GO DEEPER** Who sold more pieces of fruit?

_____

How many more?

_____ more pieces of fruit

16. **THINK SMARTER** At the city park theater, 152 people watched the morning play. Another 167 watched the afternoon play.

How many people watched the two plays? _____ people

Fill in the bubble next to each true sentence about how to solve the problem.

○ You need to regroup the tens as 1 ten and 9 ones.

○ You need to regroup the tens as 1 hundred and 1 ten.

○ You need to add 2 ones + 7 ones.

○ You need to add 1 hundred + 1 hundred + 1 hundred.

 TAKE HOME ACTIVITY • Have your child choose a new combination of two fruits on this page and find the total number of pieces of the two types of fruit.

Name _____

3-Digit Addition: Regroup Tens

 COMMON CORE STANDARD—2.NBT.B.7
Use place value understanding and properties of operations to add and subtract.

Write the sum.

1.

Hundreds	Tens	Ones
☐	☐	
1	8	7
+ 2	3	2

2.

Hundreds	Tens	Ones
☐	☐	
3	2	2
+ 3	5	6

3.

Hundreds	Tens	Ones
☐	☐	
2	8	5
+ 5	3	1

4.

4	4	5
+	3	4

5.

6	2	0
+ 2	8	8

6.

5	5	7
+ 1	8	0

Problem Solving

Solve. Write or draw to explain.

7. There are 142 blue toy cars and 293 red toy cars at the toy store. How many toy cars are there?

_____ toy cars

8. 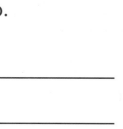 Find the sum of 362 + 265. Explain why you did or did not regroup.

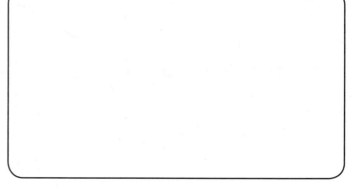

_____

_____

1. What is the sum?

$$472 + 255$$

2. Annika has 144 pennies and Yahola has 284 pennies. How many do they have altogether?

$$144 + 284$$

3. What is the sum?

$$56 + 38$$

4. What is the sum?

$$326 + 139$$

5. Francis had 8 toy cars, then his brother gave him 9 more. How many toy cars does Francis have now?

$$8 + 9 = \underline{}\ \text{cars}$$

6. What is the difference?

$$82 - 34$$

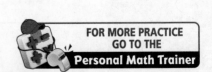

FOR MORE PRACTICE
GO TO THE
Personal Math Trainer

Addition: Regroup Ones and Tens

Essential Question How do you know when to regroup in addition?

Common Core **Number and Operations in Base Ten—2.NBT.B.7** *Also 2.NBT.B.9*
MATHEMATICAL PRACTICES
MP1, MP6, MP8

Listen and Draw Real World

Use mental math. Write the sum for each problem.

$$\begin{array}{r} 40 \\ +\ 20 \\ \hline \end{array} \qquad \begin{array}{r} 200 \\ +\ 700 \\ \hline \end{array} \qquad \begin{array}{r} 70 \\ +\ 30 \\ \hline \end{array} \qquad \begin{array}{r} 500 \\ +\ 300 \\ \hline \end{array}$$

$10 + 30 + 40 = $ _____

$100 + 400 + 200 = $ _____

$10 + 50 + 40 = $ _____

$600 + 300 = $ _____

FOR THE TEACHER • Encourage children to do these addition problems quickly. You may wish to first discuss the problems with children, noting that each problem is limited to just adding tens or just adding hundreds.

Math Talk MATHEMATICAL PRACTICES 1

Analyze Were some problems easier to solve than other problems? Explain.

Model and Draw

Sometimes you will regroup more than once in addition problems.

$$\begin{array}{r} 1\ 1 \\ 2\ 5\ 9 \\ +\ 4\ 7\ 6 \\ \hline 7\ 3\ 5 \end{array}$$

> 9 ones + 6 ones = 15 ones, or 1 ten 5 ones

> 1 ten + 5 tens + 7 tens = 13 tens, or 1 hundred 3 tens

> 1 hundred + 2 hundreds + 4 hundreds = 7 hundreds

THINK:
Are there 10 or more ones?
Are there 10 or more tens?

Share and Show

Write the sum.

1.
$$\begin{array}{r} 1\ 8\ 4 \\ +\ 3\ 2\ 9 \\ \hline \end{array}$$

2.
$$\begin{array}{r} 5\ 4\ 6 \\ +\ 2\ 7\ 8 \\ \hline \end{array}$$

3.
$$\begin{array}{r} 3\ 2\ 7 \\ +\ 3\ 5\ 3 \\ \hline \end{array}$$

4.
$$\begin{array}{r} 2\ 3\ 4 \\ +\ 1\ 5\ 2 \\ \hline \end{array}$$

✓5.
$$\begin{array}{r} 3\ 7\ 5 \\ +\ 2\ 7\ 2 \\ \hline \end{array}$$

✓6.
$$\begin{array}{r} 1\ 8\ 9 \\ +\ 6\ 2\ 3 \\ \hline \end{array}$$

Name _____

On Your Own

Write the sum.

7.
```
  5 7 4
+ 2 8 1
———————
```

8.
```
  4 1 6
+ 4 8 3
———————
```

9.
```
  3 4 6
+ 5 9 7
———————
```

10.
```
  3 6 5
+ 2 8 3
———————
```

11.
```
  6 4 7
+ 1 0 9
———————
```

12.
```
  5 4 6
+ 3 5 6
———————
```

13.
```
  3 4 8
+ 6 3 1
———————
```

14.
```
  4 5 5
+ 1 3 9
———————
```

15.
```
  5 6 3
+ 2 4 5
———————
```

16. **THINK SMARTER** Miko wrote these problems.
What are the missing digits?

```
  ▇ ▇ 6
+   4 5 ▇
—————————
  6 9 0
```

```
  6 ▇ 7
+ 2 3 ▇
—————————
  ▇ 6 2
```

TAKE HOME ACTIVITY • Have your child explain how to
solve 236 + 484.

Mid-Chapter Checkpoint

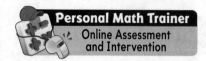

Concepts and Skills

Break apart the addends to find the sum. (2.NBT.B.7)

1. $\quad$ 567 $\longrightarrow$ _____ + _____ + _____

$\quad$ +324 $\longrightarrow$ _____ + _____ + _____

$\qquad$ _____ + _____ + _____ = _____

Write the sum. (2.NBT.B.7)

2.
$$
\begin{array}{r}
2\ 4\ 8 \\
+\ 3\ 4\ 6 \\
\hline
\end{array}
$$

3.
$$
\begin{array}{r}
6\ 3\ 7 \\
+\ 2\ 6\ 4 \\
\hline
\end{array}
$$

4.
$$
\begin{array}{r}
3\ 9\ 1 \\
+\ 5\ 3\ 7 \\
\hline
\end{array}
$$

5. **THINK SMARTER** There are 148 small sand dollars and 119 large sand dollars on the beach. How many sand dollars are on the beach? (2.NBT.B.7)

_____ sand dollars

Addition: Regroup Ones and Tens

Common Core **COMMON CORE STANDARD—2.NBT.B.7**
Use place value understanding and properties of
operations to add and subtract.

Write the sum.

1.
```
    5 4 7
  + 4 3 5
```

2.
```
    3 6 7
  + 2 8 4
```

3.
```
    4 8 5
  + 4 5 6
```

4.
```
    1 8 7
  + 3 0 6
```

5.
```
    6 4 7
  + 1 2 8
```

6.
```
    5 2 3
  + 1 7 4
```

Problem Solving

Solve. Write or draw to explain.

7. Saul and Luisa each scored 167 points
on a computer game. How many points
did they score together?

_____ points

8. **WRITE** Math Write the addition
problem for 275 plus 249 and
find the sum. Then draw quick
pictures to check your work.

_____

Lesson Check (2.NBT.B.7)

1. What is the sum?

$$348 + 272$$

2. What is the sum?

$$123 + 217$$

Spiral Review (2.OA.A.1, 2.OA.B.2, 2.NBT.B.6, 2.NBT.B.9)

3. Write an addition fact that has the same sum as 9 + 4.

10 + _____

4. What is the sum?

$$\begin{array}{r} 32 \\ 15 \\ + 46 \\ \hline \end{array}$$

5. Add 29 and 35. Draw to show the regrouping. What is the sum?

Tens	Ones

_____

6. Tom had 25 pretzels. He gave away 12 pretzels. How many pretzels does Tom have left?

25 − 12 = _____ pretzels

420 four hundred twenty

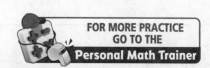

FOR MORE PRACTICE
GO TO THE
Personal Math Trainer

Name _____

Problem Solving • 3-Digit Subtraction

Essential Question How can making a model help when solving subtraction problems?

Common Core
Number and Operations in Base Ten—2.NBT.B.7
MATHEMATICAL PRACTICES
MP1, MP4, MP6

There were 436 people at the art show. 219 people went home. How many people stayed at the art show?

 Unlock the Problem

What do I need to find?

how many people
_____
stayed at the art show

What information do I need to use?

_____ people were at the art show.

Then, _____ people went home.

Show how to solve the problem.

Make a model. Then draw a quick picture of your model.

_____ people

HOME CONNECTION • Your child used a model and a quick picture to represent and solve a subtraction problem.

Make a model to solve. Then draw
a quick picture of your model.

- What do I need to find?
- What information do I need to use?

1. There are 532 pieces of art at the show. 319 pieces of art are paintings. How many pieces of art are not paintings?

_____ pieces of art

2. 245 children go to the face-painting event. 114 of the children are boys. How many of the children are girls?

_____ girls

Math Talk

MATHEMATICAL PRACTICES 6

Explain how you solved the first problem on this page.

Name _____

Make a model to solve. Then draw
a quick picture of your model.

3. There were 237 books on the
table. Miss Jackson took
126 books off the table.
How many books were still
on the table?

_____ books

4. There were 232 postcards on
the table. The children used
118 postcards. How many
postcards were not used?

_____ postcards

5. _THINK SMARTER_ 164 children and
31 adults saw the movie in the
morning. 125 children saw the
movie in the afternoon. How
many fewer children saw the
movie in the afternoon than
in the morning?

_____ fewer children

On Your Own

MATHEMATICAL PRACTICE ① Make Sense of Problems

6. There were some grapes in a bowl. Clancy's friends ate 24 of the grapes. Then there were 175 grapes in the bowl. How many grapes were in the bowl before?

_____ grapes

7. **THINK SMARTER** At Gregory's school, there are 547 boys and girls. There are 246 boys. How many girls are there?

Draw a quick picture to solve.

Circle the number that makes the sentence true.

There are

| 201 |
| 301 |
| 793 |

girls.

TAKE HOME ACTIVITY • Ask your child to choose one of the problems in this lesson and solve it in a different way.

Problem Solving • 3-Digit Subtraction

Common Core
COMMON CORE STANDARD—2.NBT.B.7
Use place value understanding and properties of operations to add and subtract.

Make a model to solve. Then draw a quick picture of your model.

1. On Saturday, 770 people went to the snack shop. On Sunday, 628 people went. How many more people went to the snack shop on Saturday than on Sunday?

_____ more people

2. There were 395 lemon ice cups at the snack shop. People bought 177 lemon ice cups. How many lemon ice cups are still at the snack shop?

_____ cups

3. There were 576 bottles of water at the snack shop. People bought 469 bottles of water. How many bottles of water are at the snack shop now?

_____ bottles

4. **WRITE** **Math** Draw quick pictures to show how to subtract 314 and 546.

Lesson Check (2.NBT.B.7)

1. There are 278 math and science books. 128 of them are math books. How many science books are there?

_____ books

2. A book has 176 pages. Mr. Roberts has read 119 pages. How many pages does he have left to read?

_____ pages

Spiral Review (2.OA.B.2, 2.NBT.B.5, 2.NBT.B.6, 2.NBT.B.7)

3. What is the sum?

$$1 + 6 + 2 = \underline{\quad}$$

4. What is the difference?

$$54 - 8 = \underline{\quad}$$

5. What is the sum?

$$\begin{array}{r} 356 \\ + \ 174 \\ \hline \end{array}$$

6. What is the sum?

$$\begin{array}{r} 22 \\ + \ 16 \\ \hline \end{array}$$

FOR MORE PRACTICE
GO TO THE
Personal Math Trainer

Name _____

3-Digit Subtraction: Regroup Tens

Essential Question When do you regroup tens in subtraction?

Common Core **Number and Operations in Base Ten—2.NBT.B.7**
MATHEMATICAL PRACTICES
MP1, MP6, MP8

Listen and Draw (Real World) (Hands On)

Use ▦ ▭ to model the problem.
Draw a quick picture to show what you did.

Hundreds	Tens	Ones

Math Talk MATHEMATICAL PRACTICES
Describe what to do when there are not enough ones to subtract from.

FOR THE TEACHER • Read the following problem and have children model it with blocks. 473 people went to the football game. 146 people were still there at the end of the game. How many people left before the end of the game? Have children draw quick pictures of their models.

$354 - 137 = ?$

Are there enough ones to subtract 7?

yes   (no)

Regroup 1 ten as 10 ones.

Hundreds	Tens	Ones
	4	14
3	5	4
− 1	3	7

Hundreds	Tens	Ones

Now there are enough ones.

Subtract the ones.

$14 - 7 = 7$

Hundreds	Tens	Ones
	4	14
3	5	4
− 1	3	7
		7

Hundreds	Tens	Ones

Subtract the tens.

$4 - 3 = 1$

Subtract the hundreds.

$3 - 1 = 2$

Hundreds	Tens	Ones
	4	14
3	5	4
− 1	3	7
2	1	7

Hundreds	Tens	Ones

Share and Show  MATH BOARD

Solve. Write the difference.

✓ 1.

Hundreds	Tens	Ones
	☐	☐
4	3	1
− 3	2	6

✓ 2.

Hundreds	Tens	Ones
	☐	☐
6	5	8
− 2	3	7

Name _____

On Your Own

Solve. Write the difference.

3.

Hundreds	Tens	Ones
	☐	☐
7	2	8
− 1	0	7

4.

Hundreds	Tens	Ones
	☐	☐
4	5	2
− 2	1	6

5.

Hundreds	Tens	Ones
	☐	☐
9	6	5
− 2	3	8

6.

Hundreds	Tens	Ones
	☐	☐
4	8	9
− 1	4	9

7. **GO DEEPER** A bookstore has
148 books about people and
136 books about places. Some
books were sold. Now there are
137 books left. How many books
were sold?

_____ books

8. **THINK SMARTER** There were
287 music books and
134 science books in the store.
After some books were sold,
there are 159 books left.
How many books were sold?

_____ books

Problem Solving • Applications   WRITE Math

MATHEMATICAL PRACTICE ① Make Sense of Problems

Solve. Write or draw to explain.

9. There are 235 whistles and 42 bells in the store. Ryan counts 128 whistles on the shelf. How many whistles are not on the shelf?

_____ whistles

Personal Math Trainer

10. **THINK SMARTER +** Dr. Jackson had 326 stamps.

He sells 107 stamps. How many stamps does he have now?

_____ stamps

Would you do these things to solve the problem?
Choose Yes or No.

Subtract 107 from 326.	○ Yes	○ No
Regroup 1 ten as 10 ones.	○ Yes	○ No
Regroup the hundreds.	○ Yes	○ No
Subtract 7 ones from 16 ones.	○ Yes	○ No
Add 26 + 10.	○ Yes	○ No

 TAKE HOME ACTIVITY • Ask your child to explain why he or she regrouped in only some of the problems in this lesson.

3-Digit Subtraction: Regroup Tens

Common Core **COMMON CORE STANDARD—2.NBT.B.7**
Use place value understanding and properties of operations to add and subtract.

Solve. Write the difference.

1.

Hundreds	Tens	Ones
	☐	☐
7	7	4
− 2	3	6

2.

Hundreds	Tens	Ones
	☐	☐
5	5	1
− 1	1	3

3.

Hundreds	Tens	Ones
	☐	☐
4	8	9
− 2	7	3

4.

Hundreds	Tens	Ones
	☐	☐
7	7	2
− 2	5	4

Problem Solving

Solve. Write or draw to explain.

5. There were 985 pencils. Some pencils were sold. Then there were 559 pencils left. How many pencils were sold?

_____ pencils

6.  WRITE Math Choose one exercise from above. Draw quick pictures to check your work.

Lesson Check (2.NBT.B.7)

1. What is the difference?

$$346 - 127$$

2. What is the difference?

$$568 - 226$$

Spiral Review (2.OA.A.1, 2.OA.C.4, 2.NBT.B.5, 2.NBT.B.7)

3. What is the difference?

$$45 - 7 = \underline{\hspace{1cm}}$$

4. Leroy has 11 cubes. Jane has 15 cubes. How many cubes do they have altogether?

_____ cubes

5. Mina puts 5 flowers in each vase. How many flowers will she put in 3 vases?

_____ flowers

6. Mr. Hill had 428 pencils. He gave away 150 pencils. How many pencils did he keep?

_____ pencils

FOR MORE PRACTICE
GO TO THE
Personal Math Trainer

Name_____

3-Digit Subtraction: Regroup Hundreds

Essential Question When do you regroup hundreds in subtraction?

Common Core **Number and Operations in Base Ten—2.NBT.B.7, 2.NBT.B.9**
MATHEMATICAL PRACTICES
MP1, MP3, MP8

Listen and Draw Real World

Draw quick pictures to show the problem.

Hundreds	Tens	Ones

Math Talk MATHEMATICAL PRACTICES I

Describe what to do when there are not enough tens to subtract from.

FOR THE TEACHER • Read the following problem and have children model it with quick pictures. The Reading Club has 349 books. 173 of the books are about animals. How many books are not about animals?

Chapter 6

four hundred thirty-three **433**

428 − 153 = ?

Subtract the ones.

8 − 3 = 5

Hundreds	Tens	Ones
4	2	8
− 1	5	3
		5

Hundreds		Tens	Ones
		‖	X

There are not enough tens to subtract from.

Regroup 1 hundred. 4 hundreds 2 tens is now 3 hundreds 12 tens.

Hundreds	Tens	Ones
3	12	
4̸	2̸	8
− 1	5	3
		5

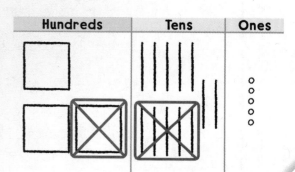

Subtract the tens.

12 − 5 = 7

Subtract the hundreds.

3 − 1 = 2

Hundreds	Tens	Ones
3	12	
4̸	2̸	8
− 1	5	3
2	7	5

Hundreds	Tens	Ones
	‖‖‖	‖
⊠	⊠	

Share and Show   MATH BOARD

Solve. Write the difference.

✓ 1.

Hundreds	Tens	Ones
4	7	8
− 3	5	6

✓ 2.

Hundreds	Tens	Ones
8	1	4
− 2	6	3

On Your Own

Solve. Write the difference.

3.

Hundreds	Tens	Ones
☐	☐	☐
6	2	9
− 4	8	2

4.

Hundreds	Tens	Ones
☐	☐	☐
9	3	6
− 1	7	3

5.

4	3	5
− 1	9	2

6.

3	8	7
−	4	7

7.

```
   5 8 8
 − 4 5 0
```

8.

```
   3 4 5
 − 2 6 3
```

MATHEMATICAL PRACTICE ③ Make Arguments

9. Choose one exercise above. Describe the subtraction that you did. Be sure to tell about the values of the digits in the numbers.

_____

_____

_____

_____

Problem Solving • Applications   WRITE ▸ Math

10. **THINK SMARTER** Sam made two towers. He used 139 blocks for the first tower. He used 276 blocks in all. For which tower did he use more blocks? _____

Explain how you solved the problem.

_____

_____

11. **THINK SMARTER** This is how many points each class scored in a math game.

Mrs. Rose
444 points

Mr. Chang
429 points

Mr. Pagano
293 points

How many more points did Mr. Chang's class score than Mr. Pagano's class? Draw a picture and explain how you found your answer.

_____ more points

 TAKE HOME ACTIVITY • Have your child explain how to find the difference for 745 − 341.

3-Digit Subtraction: Regroup Hundreds

COMMON CORE STANDARD—2.NBT.B.7
Use place value understanding and properties of operations to add and subtract.

Solve. Write the difference.

1.

Hundreds	Tens	Ones
☐	☐	☐
7	2	7
− 2	5	6

2.

Hundreds	Tens	Ones
☐	☐	☐
9	6	7
− 1	5	3

3.

6	3	9
− 4	7	2

4.

4	4	8
− 3	6	3

Problem Solving Real World

Solve. Write or draw to explain.

5. There were 537 people in the parade. 254 of these people were playing an instrument. How many people were not playing an instrument?

_____ people

6. 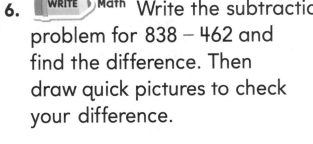 WRITE ▸ Math Write the subtraction problem for 838 − 462 and find the difference. Then draw quick pictures to check your difference.

_____

Lesson Check (2.NBT.B.7)

1. What is the difference?

$$\begin{array}{r} 538 \\ -135 \\ \hline \end{array}$$

2. What is the difference?

$$\begin{array}{r} 218 \\ -126 \\ \hline \end{array}$$

Spiral Review (2.OA.B.2, 2.NBT.B.5, 2.NBT.B.6, 2.NBT.B.7)

3. What is the difference?

$$52 - 15 = \underline{\hspace{1cm}}$$

4. Wallace has 8 crayons and Alma has 7. How many do the have together?

$$8 + 7 = \underline{\hspace{1cm}} \text{ crayons}$$

5. What is the sum?

$$\begin{array}{r} 47 \\ +26 \\ \hline \end{array}$$

6. Mrs. Lin's class read 392 books in February. Mr. Hook's class read 173 books. How many more books did Mrs. Lin's class read?

$$\begin{array}{r} 392 \\ -173 \\ \hline \end{array}$$

_____ books

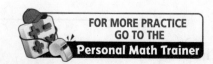

FOR MORE PRACTICE
GO TO THE
Personal Math Trainer

Name _____

Subtraction: Regroup Hundreds and Tens

Common Core Operations and Algebraic Thinking—2.NBT.B.7 *Also 2.NBT.B.8*
MATHEMATICAL PRACTICES
MP1, MP6, MP8

Essential Question How do you know when to regroup in subtraction?

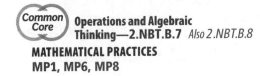

Listen and Draw *Real World*

Use mental math. Write the difference for each problem.

$$\begin{array}{r} 50 \\ -\ 20 \\ \hline \end{array} \qquad \begin{array}{r} 600 \\ -\ 400 \\ \hline \end{array} \qquad \begin{array}{r} 80 \\ -\ 30 \\ \hline \end{array} \qquad \begin{array}{r} 900 \\ -\ 300 \\ \hline \end{array}$$

$90 - 40 =$ _____

$700 - 500 =$ _____

$70 - 60 =$ _____

$800 - 300 =$ _____

Math Talk MATHEMATICAL PRACTICES 6

Were some problems easier to solve than other problems? **Explain.**

FOR THE TEACHER • Encourage children to do these subtraction problems quickly. You may wish to first discuss the problems with children, noting that each problem is limited to just subtracting tens or just subtracting hundreds.

Model and Draw

Sometimes you will regroup more than once in subtraction problems.

```
    11
  6  7  15
  7  2  5
-  3  4  9
─────────
  3  7  6
```

Regroup 2 tens 5 ones as 1 ten 15 ones. Subtract the ones.

Regroup 7 hundreds 1 ten as 6 hundreds 11 tens. Subtract the tens.

Subtract the hundreds.

Share and Show MATH BOARD

Solve. Write the difference.

1.
```
    4  2  1
 -  1  3  8
```

2.
```
    2  7  4
 -  1  8  2
```

3.
```
    5  4  6
 -  2  6  7
```

4.
```
    8  5  9
 -     5  7
```

☑5.
```
    7  4  7
 -  1  5  9
```

☑6.
```
    9  3  8
 -  3  7  0
```

Name _____

On Your Own

Solve. Write the difference.

7.
```
    3  4  2
 -  1  3  8
 _____
```

8.
```
    4  6  3
 -  2  8  1
 _____
```

9.
```
    8  5  5
 -  4  9  7
 _____
```

10.
```
    6  5  7
 -  3  8  4
 _____
```

11.
```
    5  2  1
 -  1  4  6
 _____
```

12.
```
    7  5  8
 -  5  3  7
 _____
```

13.
```
    5  4  2
 -  1  6  8
 _____
```

14.
```
    8  2  3
 -  6  7  3
 _____
```

15.
```
    9  4  7
 -  5  7  9
 _____
```

16. **THINK SMARTER** Alex wrote these problems.
What are the missing digits?

```
        4  15
    9  ▢  ▢
 -  6  2  8
 _____
    3  2  7
```

```
    7  13
    ▢  ▢  7
 -  1  5  ▢
 _____
    6  8  1
```

Problem Solving • Applications WRITE  Math

17. **GO DEEPER** This is how Walter found the difference for 617 — 350.

Find the difference for 843 — 270 using Walter's way.

350
400 } + 50
600 } + 200
617 } + 17

(267)

18. **MATHEMATICAL PRACTICE ①** **Analyze** There are 471 children at Caleb's school. 256 children ride buses to get to school.

How many children do not ride buses to get to school?

_____ children

19. **THINK SMARTER** Mrs. Herrell had 427 pinecones. She gave 249 pinecones to her children.

How many pinecones does she still have?

_____ pinecones

 TAKE HOME ACTIVITY • Ask your child to find the difference when subtracting 182 from 477.

Subtraction: Regroup Hundreds and Tens

Common Core **COMMON CORE STANDARD—2.NBT.B.7**
Use place value understanding and properties of operations to add and subtract.

Solve. Write the difference.

1.
```
   8  1  6
-  3  4  5
```

2.
```
   9  3  2
-  1  6  3
```

3.
```
   7  9  6
-  4  6  8
```

Problem Solving

Solve.

4. Mia's coloring book has 432 pages.
She has already colored 178 pages.
How many pages in the book are
left to color?

_____ pages

5. **WRITE** **Math** Draw quick
pictures to show how to
subtract 546 from 735.

Lesson Check (2.NBT.B.7)

1. What is the difference?

$$\begin{array}{r} 349 \\ -\ 187 \\ \hline \end{array}$$

2. What is the difference?

$$\begin{array}{r} 336 \\ -\ 178 \\ \hline \end{array}$$

Spiral Review (2.OA.A.1, 2.OA.B.2, 2.NBT.B.5, 2.NBT.B.7)

3. What is the sum?

$$\begin{array}{r} 246 \\ +\ 533 \\ \hline \end{array}$$

4. What is the difference?

$$\begin{array}{r} 38 \\ -\ 14 \\ \hline \end{array}$$

5. What is the difference?

$$17 - 9 = \underline{\hspace{1cm}}$$

6. Lisa had 15 daisies.
She gave away 7 daisies.
Then she found 3 more daisies.
How many daisies does Lisa
have now?

_____ daisies

FOR MORE PRACTICE
GO TO THE
Personal Math Trainer

Name _____

Regrouping with Zeros

Essential Question How do you regroup when there are zeros in the number you start with?

Common Core
Number and Operations in Base Ten—2.NBT.B.7
MATHEMATICAL PRACTICES
MP1, MP6, MP8

Listen and Draw Real World

Draw or write to show how you solved the problem.

Math Talk

MATHEMATICAL PRACTICES 1
Describe another way that you could solve the problem.

FOR THE TEACHER • Read the following problem and have children solve. Mr. Sanchez made 403 cookies. He sold 159 cookies. How many cookies does Mr. Sanchez have now? Encourage children to discuss and show different ways to solve the problem.

Chapter 6

Model and Draw

Ms. Dean has a book with 504 pages in it. She has read 178 pages so far. How many more pages does she still have to read?

$$\begin{array}{r} 5\ 0\ 4 \\ -\ 1\ 7\ 8 \\ \hline \end{array}$$

Step 1 There are not enough ones to subtract from.

Since there are 0 tens, regroup 5 hundreds as 4 hundreds 10 tens.

$$\begin{array}{r} \overset{4}{\cancel{5}}\ \overset{10}{\cancel{0}}\ 4 \\ -\ 1\ 7\ 8 \\ \hline \end{array}$$

Step 2 Next, regroup 10 tens 4 ones as 9 tens 14 ones.

Now there are enough ones to subtract from.

$$14 - 8 = 6$$

$$\begin{array}{r} \overset{4}{\cancel{5}}\ \overset{\overset{9}{\cancel{10}}}{\cancel{0}}\ \overset{14}{\cancel{4}} \\ -\ 1\ 7\ 8 \\ \hline 6 \end{array}$$

Step 3 Subtract the tens.

$$9 - 7 = 2$$

Subtract the hundreds.

$$4 - 1 = 3$$

$$\begin{array}{r} \overset{4}{\cancel{5}}\ \overset{\overset{9}{\cancel{10}}}{\cancel{0}}\ \overset{14}{\cancel{4}} \\ -\ 1\ 7\ 8 \\ \hline 3\ 2\ 6 \end{array}$$

Share and Show

Solve. Write the difference.

1.
$$\begin{array}{r} 3\ 0\ 8 \\ -\ 2\ 5\ 9 \\ \hline \end{array}$$

☑ 2.
$$\begin{array}{r} 7\ 5\ 5 \\ -\ 4\ 3\ 8 \\ \hline \end{array}$$

☑ 3.
$$\begin{array}{r} 8\ 0\ 1 \\ -\ 3\ 7\ 5 \\ \hline \end{array}$$

Name _____

On Your Own

Solve. Write the difference.

4.
```
    5 6 3
  - 1 8 2
```

5.
```
    9 0 4
  - 5 6 8
```

6.
```
    7 0 5
  - 2 3 1
```

7.
```
    6 0 3
  - 3 2 8
```

8.
```
    4 4 2
  - 2 3 8
```

9.
```
    9 0 1
  - 6 7 5
```

10.
```
    7 0 2
  - 4 2 6
```

11.
```
    6 8 4
  - 2 1 9
```

12.
```
    4 7 9
  - 1 3 7
```

13. **THINK SMARTER** Miguel has 125 more baseball cards than Chad. Miguel has 405 baseball cards. How many baseball cards does Chad have?

_____ baseball cards

Problem Solving • Applications  WRITE Math

14. **MATHEMATICAL PRACTICE ①** **Analyze** Claire has 250 pennies. Some are in a box and some are in her bank. There are more than 100 pennies in each place. How many pennies could be in each place?

_____ pennies in a box

_____ pennies in her bank

Explain how you solved the problem.

_____

_____

_____

_____

15. **THINK SMARTER** There are 404 people at the baseball game. 273 people are fans of the blue team. The rest are fans of the red team. How many people are fans of the red team?

Does the sentence describe how to find the answer? Choose Yes or No.

Regroup 1 ten as 14 ones.   ○ Yes   ○ No

Regroup 1 hundred as 10 tens.   ○ Yes   ○ No

Subtract 3 ones from 4 ones.   ○ Yes   ○ No

Subtract 2 hundreds from 4 hundreds.   ○ Yes   ○ No

There are _____ fans of the red team.

 TAKE HOME ACTIVITY • Ask your child to explain how he or she solved one of the problems in this lesson.

Regrouping with Zeros

Common Core   **COMMON CORE STANDARD—2.NBT.B.7**
Use place value understanding and properties of operations to add and subtract.

Solve. Write the difference.

1.
```
  8 0 6
- 3 4 5
```

2.
```
  9 0 2
- 7 8 3
```

3.
```
  7 9 4
- 2 6 8
```

4.
```
  6 8 7
- 1 4 4
```

5.
```
  5 0 5
- 1 6 7
```

6.
```
  3 0 7
- 1 5 4
```

Problem Solving Real World

Solve.

7. There are 303 students.
   There are 147 girls.
   How many boys are there?   _____ boys

8. **WRITE** ▸ Math Write the
   subtraction problem 604 − 357.
   Describe how you will subtract
   to find the difference.

   _____

   _____

I. What is the difference?

$$\begin{array}{r} 301 \\ -\ 187 \\ \hline \end{array}$$

2. What is the difference?

$$\begin{array}{r} 406 \\ -\ 268 \\ \hline \end{array}$$

Spiral Review (2.OA.B.2, 2.NBT.B.5, 2.NBT.B.7)

3. What is the sum?

$$\begin{array}{r} 35 \\ +\ 79 \\ \hline \end{array}$$

4. There are 555 students at Roosevelt Elementary School and 282 students at Jefferson Elementary. How many students are at the two schools altogether?

$$\begin{array}{r} 555 \\ +\ 282 \\ \hline \end{array}$$

_____ students

5. What is the difference?

$$10 - 2 = \underline{\quad}$$

6. Gabriel's goal is to read 43 books this year. So far he has read 11 books. How many books does he have left to meet his goal?

$$\begin{array}{r} 43 \\ -\ 11 \\ \hline \end{array}$$

_____ books

**FOR MORE PRACTICE
GO TO THE
Personal Math Trainer**

Name _____

 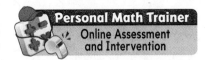
1. Mr. Kent had 948 craft sticks. His art class used 356 craft sticks. How many craft sticks does Mr. Kent have now?

   _____ craft sticks

2. At the library, there are 668 books and magazines. There are 565 books at the library. How many magazines are there?

   Circle the number that makes the sentence true.

   There are
   | 13 |
   | 103 |
   | 1,233 |
   magazines.

3. There are 176 girls and 241 boys at school. How many children are at school?

   $176 \longrightarrow 100 + 70 + 6$
   $+ 241 \longrightarrow 200 + 40 + 1$

   Select one number from each column to solve the problem.

Hundreds	Tens	Ones
○ 2	○ 1	○ 3
○ 3	○ 3	○ 5
○ 4	○ 4	○ 7

4. **THINK SMARTER +** Anna wants to add 246 and 132.

Help Anna solve this problem. Draw quick pictures. Write how many hundreds, tens, and ones in all. Write the number.

Hundreds	Tens	Ones

_____ hundreds _____ tens _____ ones

_____

---

5. Mrs. Preston had 513 leaves. She gave 274 leaves to her students. How many leaves does she still have? Draw to show how you found your answer.

_____ leaves

---

6. A farmer has 112 pecan trees and 97 walnut trees. How many more pecan trees does the farmer have than walnut trees?

Fill in the bubble next to all the sentences that describe what you would do.

○ I would regroup the hundreds.

○ I would add 12 + 97.

○ I would subtract 7 ones from 12 ones.

○ I would regroup the tens.

452 four hundred fifty-two

7. Amy has 408 beads. She gives 322 beads to her sister. How many beads does Amy have now?

Does the sentence describe how to find the answer? Choose Yes or No.

Regroup 1 ten as 18 ones.	○ Yes	○ No
Regroup 1 hundred as 10 tens.	○ Yes	○ No
Subtract 2 tens from 10 tens.	○ Yes	○ No

Amy has _____ beads.

---

8. **GO DEEPER** Raul used this method to find the sum 427 + 316.

```
   427
 + 316
   700

    30
 +  13
   743
```

Use Raul's method to find this sum.

```
   229
 + 313
```

Describe how Raul solves addition problems.

_____

_____

9. Sally scores 381 points in a game. Ty scores 262 points.
   How many more points does Sally score than Ty?

   ○ 121   ○ 643   ○ 129   ○ 119

10. Use the numbers on the tiles to solve the problem.

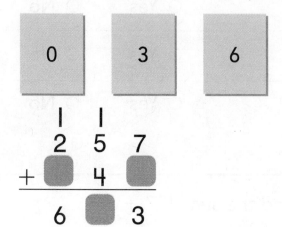

    Describe how you solved the problem.

    _____

    _____

    _____

    _____